DIGESTION, ABSORPTION AND USE OF FOOD

MUHAMMAD LAMIN JANNEH

To order additional copies of this book, contact:
Xlibris LLC
1-888-795-4274
www.Xlibris.com
Orders@Xlibris.com
589187

To order additional copies of this book, contact:
Xlibris LLC
1-888-795-4274
www.Xlibris.com
Orders@Xlibris.com
589187

To order additional copies of this book, contact:
Xlibris LLC
1-888-795-4274
www.Xlibris.com
Orders@Xlibris.com
589187

First Edition 2014.—Xlibris LLC US; 1663 Liberty Drive Bloomington, IN 47403

To order additional copies of this book, contact:
Xlibris LLC
1-888-795-4274
www.Xlibris.com
Orders@Xlibris.com
589187

Preface

The focus of this book on the digestive system is to provide resources and ideas to help students and non-students of all kinds in their academic quest. Furthermore, the goal of this book is to appeal to a wide variety of students whose desire is to study and take various exams in Biology. Therefore, as a Certified Teacher (Trained Teacher), I found it necessary and appropriate to prepare educational material like this book to benefit the population in general, and in particular students in grades 6 through grade 12 in the new Gambian Educational System and other various exams in Biology around the world.

TC and HTC students in The Gambia will find this book useful during the course of their studies at The Gambia College School of Education and in their teaching careers in schools or in various professions in science generally. Personally, I was studying Mathematics with Physics as a minor for my HTC at The Gambia College School of Education when I came across Biology.

M.L. JANNEH

Contents

Introduction

We all depend on eating for our daily survival. Eating is the process of taking food into the mouth, chewing it, and then swallowing it down into the stomach. If we go too long without food, we become hungry. This feeling tells us that, it is time to eat. Food is first broken into smaller parts, before it will be absorbed by the body. This process of breaking down food into smaller pieces, is called digestion. The food is dissolved, and then it is absorbed into the blood stream. The blood takes the broken down food to all parts of the body, such as the muscles, brain, heart, and kidneys.

The Alimentary Canal

The Alimentary Canal is a tube, that runs through the body from the mouth to the anus, in which food travels. Along this path, food is broken down, and then absorbed by the body.

A DIAGRAM OF The ALIMENTARY CANAL

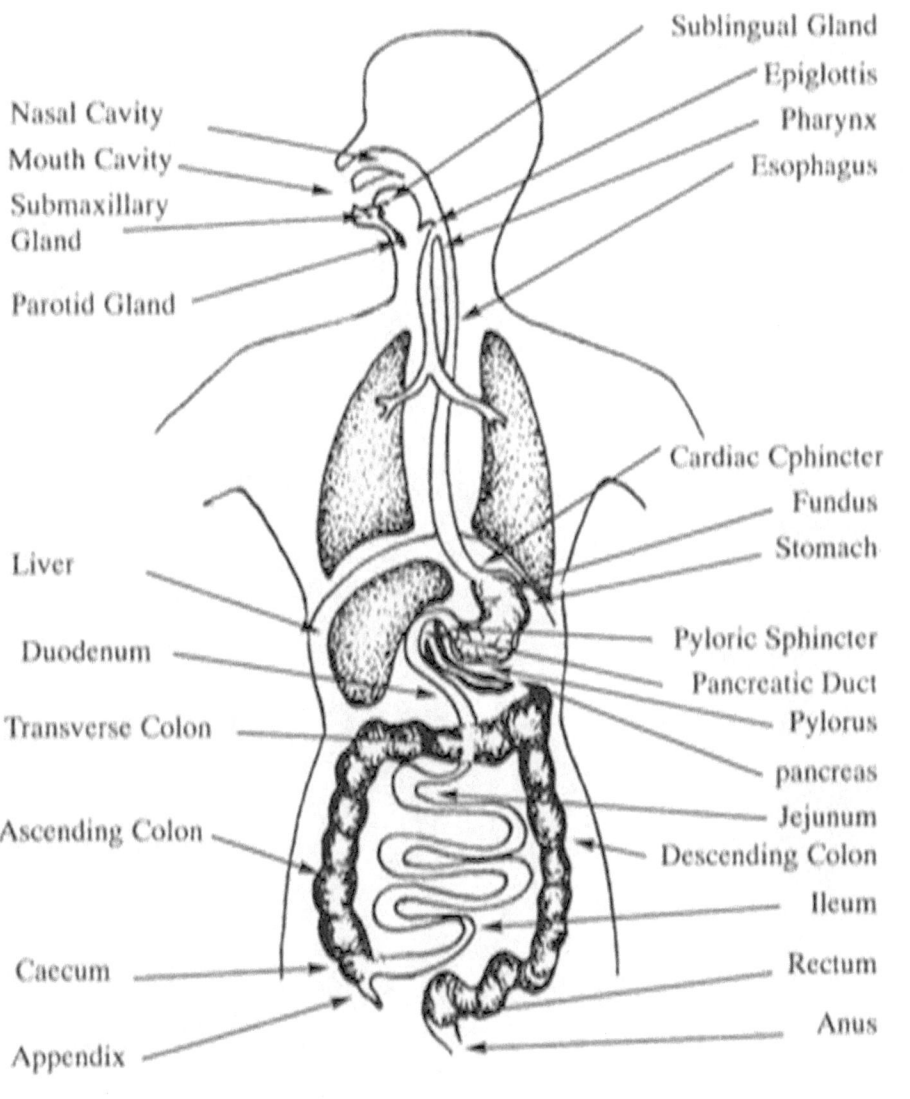

Sublingual Gland

Epiglottis

Pharynx

Esophagus

Nasal Cavity

Mouth Cavity

Submaxillary Gland

Parotid Gland

Cardiac Cphincter

Fundus

Stomach

Liver

Duodenum

Pyloric Sphincter

Pancreatic Duct

Pylorus

Transverse Colon

pancreas

Jejunum

Ascending Colon

Descending Colon

Ileum

Caecum

Rectum

Anus

Appendix

PERISTALSIS

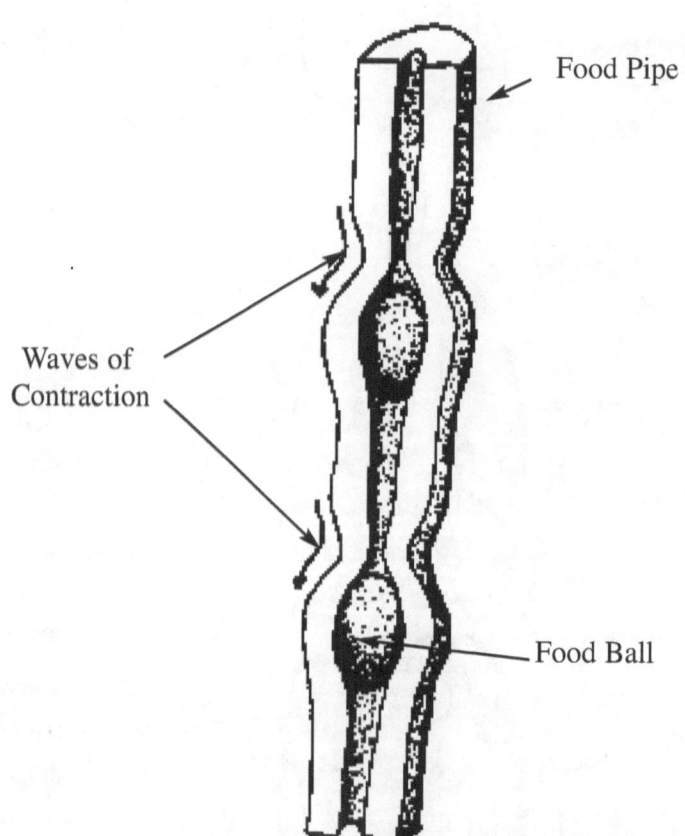

Food Pipe

Waves of
Contraction

Food Ball

**A LONGITUDINAL CROSS SECTION
OF OESOPHAGUS**

Structure

The Structure Of The Alimentary Canal

Inside the Alimentary Canal, there is a lining of layered cells called the Epithelium. The movement of the food constantly rubs the cells away, so they are always being replaced by new ones. Some of the cells along the Canal's wall, produce mucus. Mucus is a slimy liquid, which lubricates the walls of the Alimentary Canal, and also acts as a protective coating. Mucus also protects the digestive enzymes produced in the Alimentary Canal, so that food passes through more easily.
Mucus is also produced outside of the Alimentary Canal, by different glands. These glands, use tubes called ducts, to transport mucus enzymes. Two of these glands, are the Pancreas and the Salivary Glands.

Peristalsis

This is a movement of food down the food pipe or Esophagus. Food moves in the food pipe or esophagus, when it is swallowed. This movement of food in the food pipe, is called Peristalsis.

Waves of contraction along the esophagus, pushes the food down to the stomach.

DIGESTION

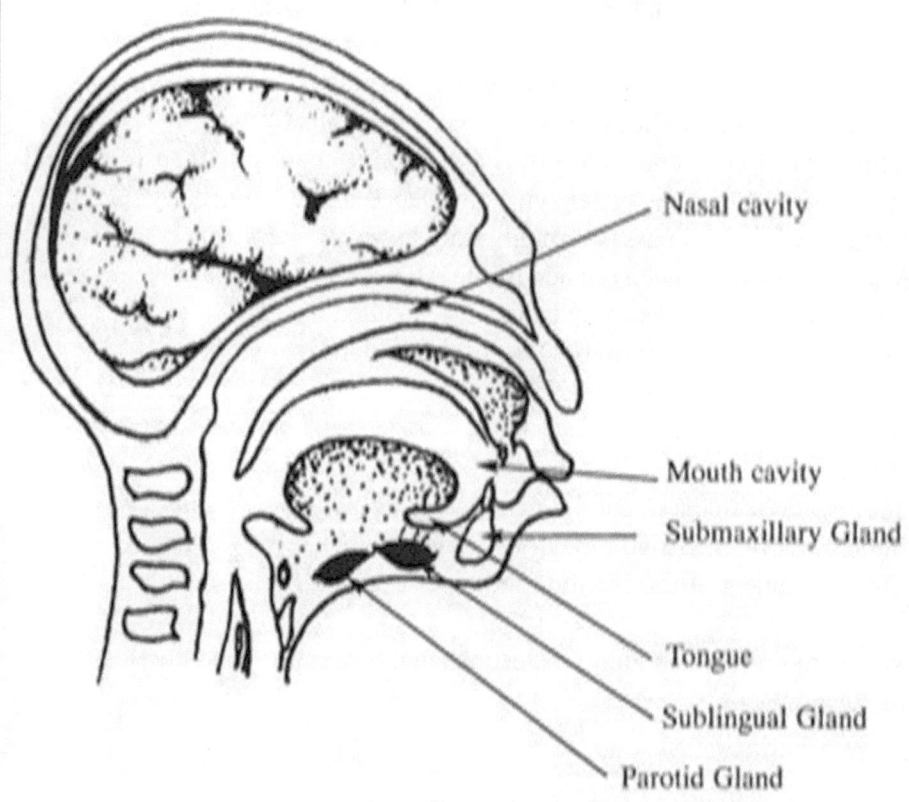

Nasal cavity

Mouth cavity

Submaxillary Gland

Tongue

Sublingual Gland

Parotid Gland

A HUMAN HEAD

Digestion

Digestion

Digestion is the breaking down of large pieces of food into smaller particles. The particles are so small that, they can go through the epithelium of the alimentary canal, through the walls of the blood vessels, into the blood. Chemicals called enzymes, help in the breakdown of the food.

Enzymes

Enzymes are chemicals, which helps to break down and dissolve food quickly.

Example of food breakdown: -

All the solid starch in foods like yam and cassaver, are broken down into glucose, which is soluble in water. Glucose is a smaller molecule than starch. The solid proteins in meat, egg, and beans are changed into smaller soluble substances, called amino acids. Fat molecules are changed into two soluble smaller substances, called glycerol and fatty acids.

This breakdown of food inside the alimentary canal, take place in different parts of the alimentary canal.

A DIAGRAMMATIC PROCESS OF INGESTION

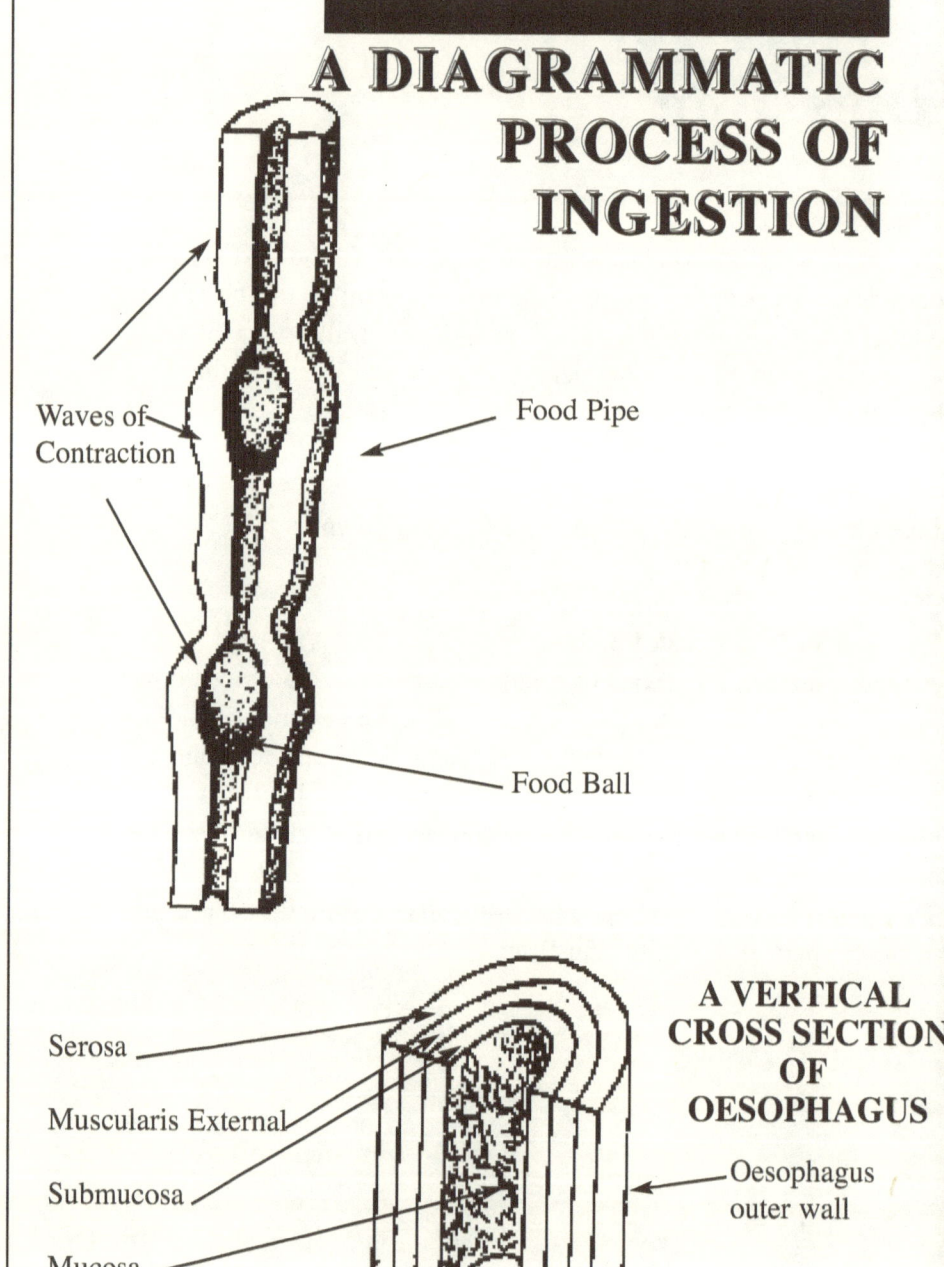

Waves of Contraction

Food Pipe

Food Ball

Serosa

Muscularis External

Submucosa

Mucosa

A VERTICAL CROSS SECTION OF OESOPHAGUS

Oesophagus outer wall

Ingestion

Stage One - Ingestion, "Digestion in the mouth"
Ingestion is the process of taking food into the mouth, where food is chewed and mixed with saliva. Saliva is a digestive juice which is produced by the three pairs of glands in the mouth. These glands have ducts, that opens into the mouth from which saliva is excreted. The saliva then will be mixed with the food, to break it down.

The glands in the mouth are as follows: *(refer to the Diagram on Page 12)*
       1. Parotid gland - The largest of the three glands located beneath the ear towards the front of the face.
       2. submaxillary gland - This gland lies along each side of the lower jaw.
       3. Sublingual gland - This gland is found underneath the tongue.

All these glands produce a digestive juice called, saliva in the mouth. Saliva mixes with the chewed food before the food moves into the esophagus. When food is mixed with the saliva in the mouth, and after the digestion of starch has already taken place in the food, the food is then moved into the back portion of the mouth, and then swallowed into the esophagus, then into the stomach

A DIAGRAM OF A HUMAN INNER STOMACH

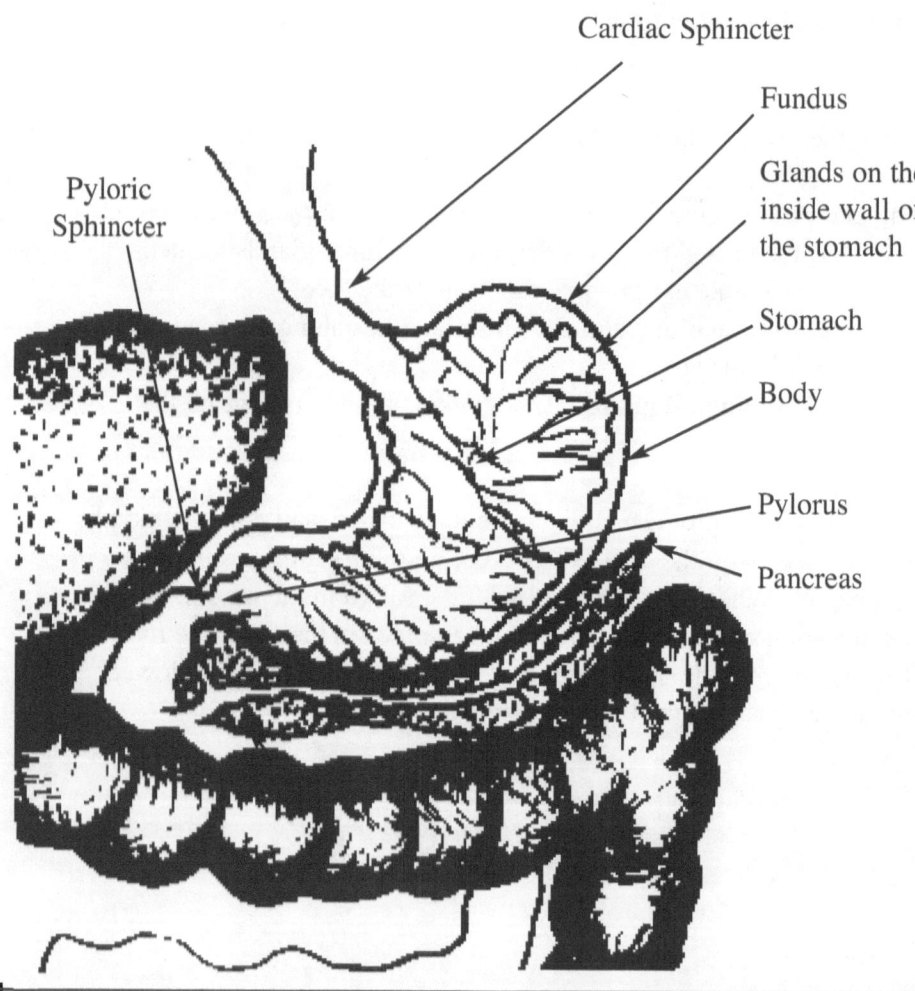

Cardiac Sphincter

Fundus

Glands on the inside wall of the stomach

Stomach

Body

Pylorus

Pancreas

Pyloric Sphincter

The Stomach

Stage Two - Digestion Inside the Stomach.

The stomach is a medium sized sac, with flexible walls that expands when food goes inside it. The main function of the stomach, is to store the food that is swallowed, change it into a liquid, and release it in small quantities, at intervals, to the rest of the alimentary canal.

The pyloric sphincter is a smooth muscle, forming a circle at the lower end of the stomach, which stops solid pieces of food, from passing through into the small intestine, or duodendum.

The cardiac sphincter is a smooth muscle, forming a circle at the upper end of the stomach, which allows food to pass through, from the esophagus into the stomach.

The stomach walls have glands, which produce a juice called, Gastric Juice. This gastric juice contains an enzyme called, pepsinogen which is used in the stomach to break down proteins into peptides and peptones. The stomach walls also produce an acid called, hydrochloric acid. Hydrochloric acid helps pepsin to work inside the stomach, and kills many of the bacteria taken into the stomach with the food.

Every 20 seconds, the peristilsic movements of the stomach mixes up the food, and gastric juices into a creamy liquid. The time that the food spends in the stomach, depends on what type of food it is. Food like porridge, may pass through the stomach in less than an hour, but foods containing protein and fat like beef, may take two hours.

A DIAGRAM OF A SMALL INTESTINE

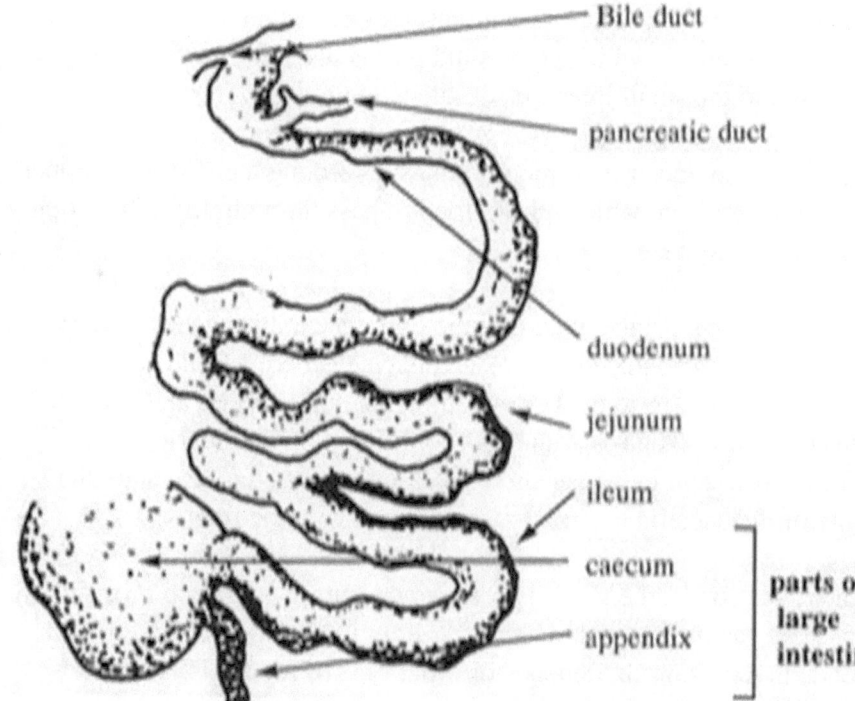

Bile duct

pancreatic duct

duodenum

jejunum

ileum

caecum

appendix

parts of large intestine

Small Intestine

After all the necessary chemical processes for digestion that must take place inside the stomach have been completed, the pyloric sphincter lets the liquid products of the digested food ball, a little at a time, into the front part of the small intestine called, the duodenum.

Stage Three - Digestion in the Small Intestine.
The pancreas is a digestive gland, that lies underneath the stomach. It produces different types of enzymes, that act on all classes of food.

> **1. Trypsin** - This enzyme breaks down proteins and peptides, into amino acids.
> **2. Amalyse** - This enzyme acts on starch, and changes it to maltose.
> **3. Lipase** - This enzyme acts on fat and changes it into fatty acids and glycerol.

The digestive juice that comes from the pancreas is called pancreatic juice. Pancreatic juice and bile, which is produced by the liver, are both released into the duodenum, to act on the food there.
Pancreatic juice is alkaline, and partly neutralizes the acidic liquids in the stomach, because the pancreas does not work well in acidic conditions.
Bile is a green, watery fluid produced by the liver, and stored in the gall bladder. It is passes into the duodenum by the bile duct, or tube. It contains no enzymes, but it is green in color, which is caused by bile pigments, formed by the breakdown of hemoglobin in the liver.
Bile contains bile salts, which acts on fats, and breaks them up into small drops, to be more easily digested by lipase.

INTESTINAL WALL

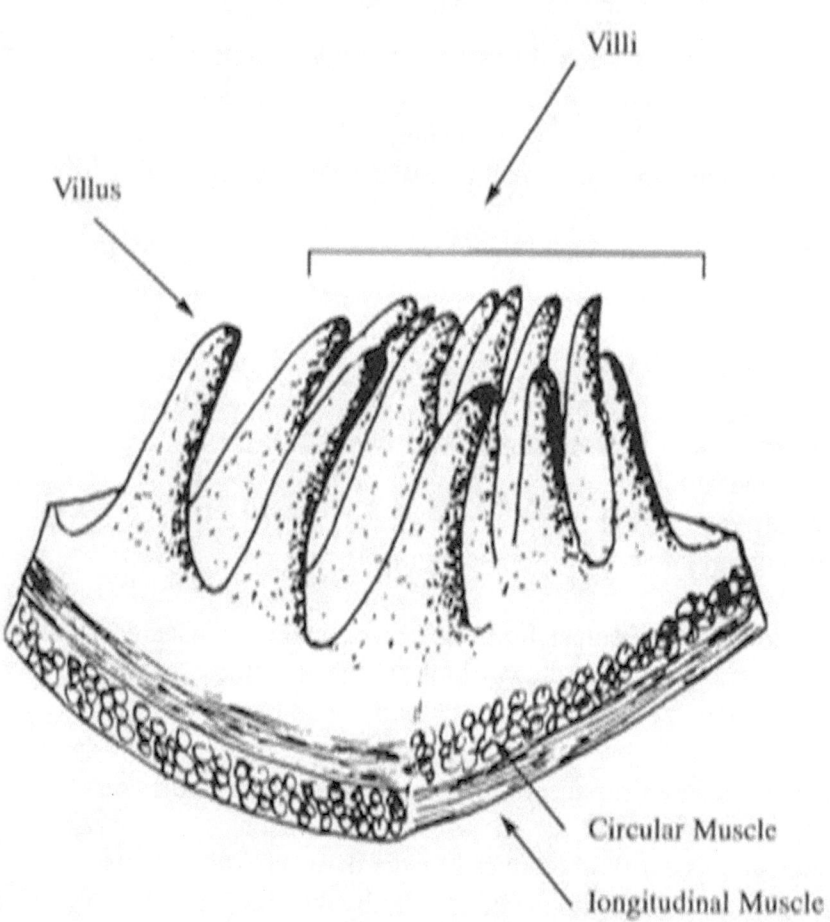

Villi

Villus

Circular Muscle

longitudinal Muscle

**INTERNAL SECTION
OF A SMALL
INTESTINE**

Final Products

All the digestible materials have now changed to soluble substances, which can pass through the wall of the intestine, and into the blood stream.

The final products of digestion are:
>   1. **Food** - Final products or nutrients
>   2. **Starch** - Glucose
>   3. **Protein**s - Amino Acids
>   4. **Fats** - Fatty Acids and Glycerol

ILEUM

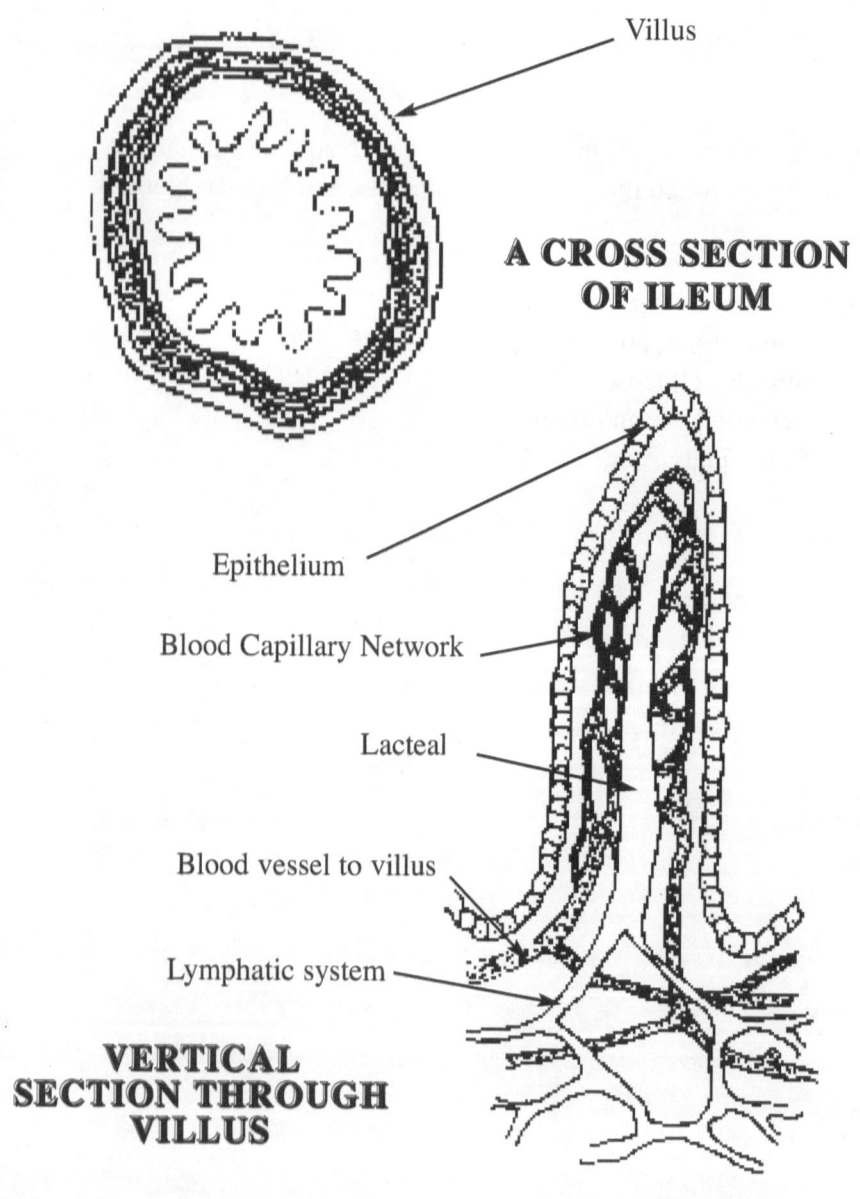

Villus

**A CROSS SECTION
OF ILEUM**

Epithelium

Blood Capillary Network

Lacteal

Blood vessel to villus

Lymphatic system

**VERTICAL
SECTION THROUGH
VILLUS**

Absorption

Stage Four - Absorption
In the small intestine, nearly all absorption of digested food takes place.
Absorption occurs easily, because of the following reasons:

1. The small intestine is long, and has
a large absorbing surface area.

2. The walls of the small intestine have
thousands of projections called villi.

3. The tiny epithelium on the walls of the small intestine are thin,
and the fluids can pass rapidly through them.

4. Inside the villus on the walls of the small intestine, there is a
network of blood capillaries in each villus for absorption.

Small molecules of the digested food like glucose and amino acids are passed through the epithelium and the capillary walls, and then enters into the blood stream. They are then carried away in the capillaries, which formed veins. These veins then come together and form one large vein called, the hepatic portal vein. Veins such as these carry all the blood from the intestine to the liver, which stores and changes any of the undigested products. When the digested food is replaced from the liver, it enters into the general blood circulation. Some of the fatty acids and glycerol from the digestion of fats, enter the blood capillaries of the villi. However, a large proportion of the fatty acids and glycerol, may be combined to form fats again in the walls of the intestine. These fats then passes into the lacteal. The fluid in the lacteal flow into the lymphatic system which empties its contents into the blood stream. Some of the materials that are not digested in the small intestine, passes into the large intestine.

A DIAGRAM OF A LARGE INTESTINE

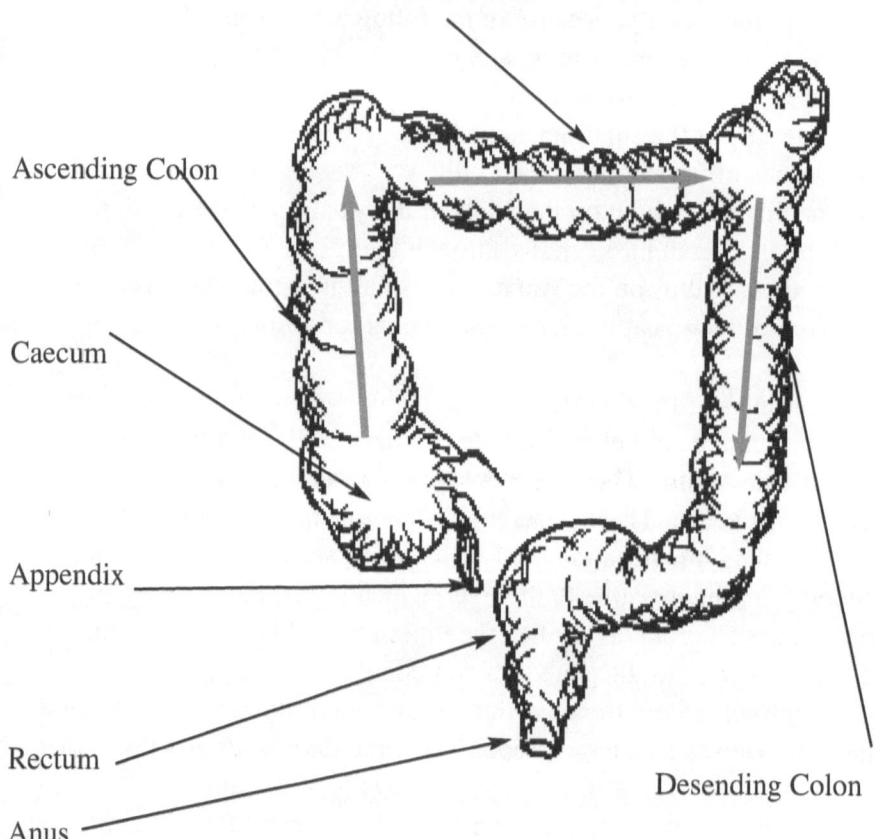

Transverse Colon

Ascending Colon

Caecum

Appendix

Rectum

Anus

Desending Colon

Large Intestine

Stage Five - Digestion in the Large Intestine
The remaining material that goes into the large intestine from the small intestine, consists of water with undigested matter, largely cellulose and vegetable fibers, mucus, and dead cells from the alimentary canal. The large intestine produces no enzymes, but the colon contains bacteria that digests part of the fiber to form fatty acids, that the colon can absorb.

In the large intestine, bile salts are absorbed and returned to the liver by the blood system. The colon absorbs much of the water from the undigested residue of materials. Semi-solid waste or feces is passed into the rectum by peristalsis and is expelled in intervals through the anus. These waste materials may spend anywhere from 12 to 24 hours in the intestine. This act of releasing feces through the anus, is called defecation. The appendix has no use in human beings, but rabbits use it for cellulose digestion. A body part that has no use, is said to be vestigial.

A DIAGRAM OF ALIMENTARY CANAL

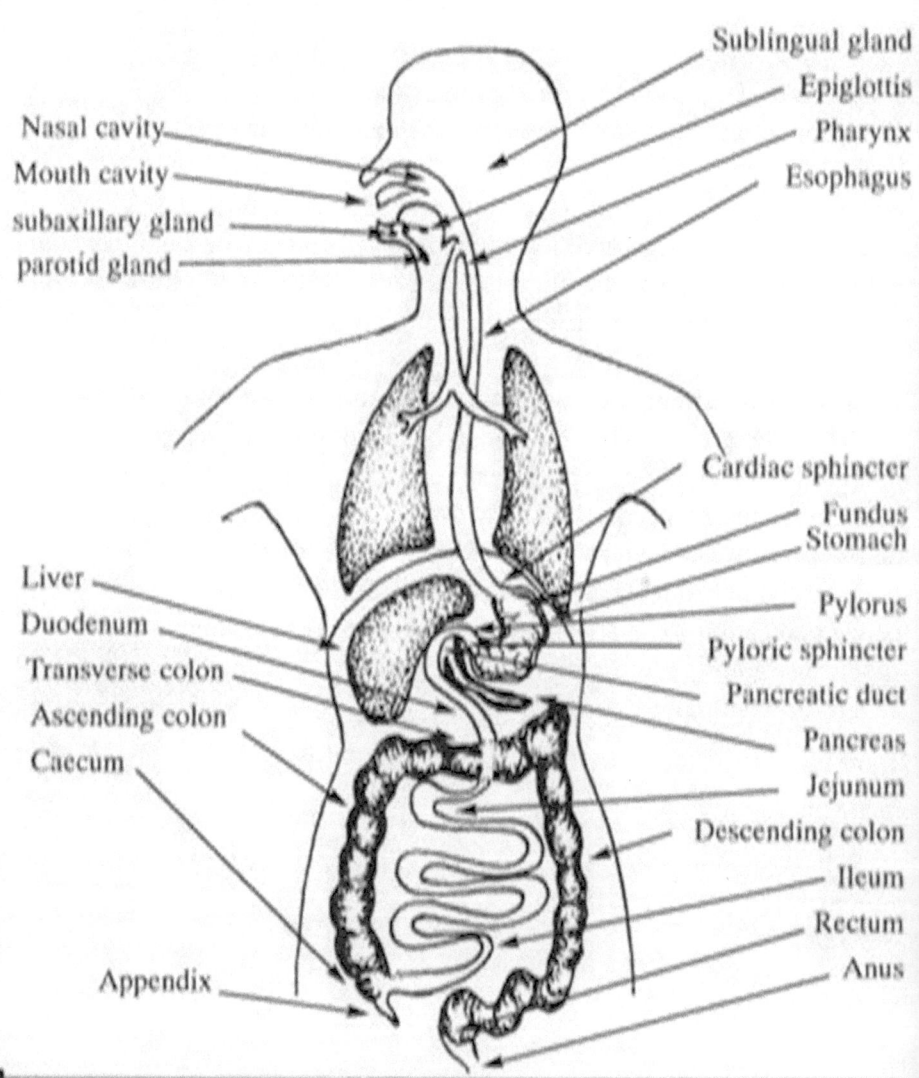

Sublingual gland
Epiglottis
Pharynx
Esophagus

Nasal cavity
Mouth cavity
subaxillary gland
parotid gland

Cardiac sphincter
Fundus
Stomach

Liver
Duodenum
Transverse colon
Ascending colon
Caecum

Pylorus
Pyloric sphincter
Pancreatic duct
Pancreas
Jejunum
Descending colon
Ileum
Rectum
Anus

Appendix

Food Uses

Uses of Digested Food
(glucose, fats and amino acids)

These products of digestion, are carried around by the blood stream to all parts of the body. From the blood, cells absorb and use the products of digestion.

GLUCOSE
During respiration in the cells, glucose is oxidized (broken down) to form carbon dioxide and water. This reaction provides energy to divide the many chemical processes in the cells which result in the building up of proteins and the contraction of muscles, for example.

FATS
These are present in cell membranes, and other cell structures. Any fats not used for growth or maintenance, are oxidized into carbon dioxide and water, providing energy for the vital processes of the cell.

AMINO ACIDS
These are absorbed by the cells and then built up into proteins. These proteins may form structures, such as the cell membrane, or they may become enzymes which control the chemical activity within the cell.

PANCREAS AND A LIVER

Pancreas

To the Liver and Gall Bladder

Pancreatic duct

A DIAGRAM OF A HUMAN PANCREAS

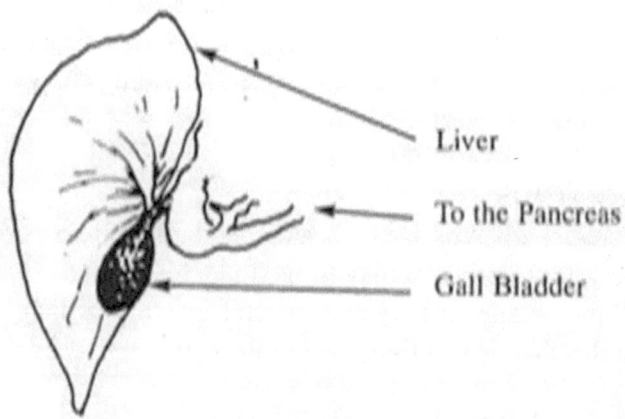

Liver

To the Pancreas

Gall Bladder

A DIAGRAM OF A HUMAN LIVER

Digestion

Use of the Liver

The liver is a red organ, which lies above the stomach, and plays a vital role in digestion. All blood from the blood vessels of the alimentary canal, passes through the liver, which adjusts the blood before releasing it into general circulation.

1. Regulation of Blood Sugar

After eating food, the liver removes excess glucose from the blood, and stores it as glycogen. In the periods between taking food, when the glucose concentration in the blood starts to fall, the liver changes some of its stored glycogen into glucose, and takes it into the blood stream. This maintains the concentration of sugar at a steady level.

2. De-amination

The amino acids, which are not needed for making proteins, are changed to glycogen in the liver. During this process, the nitrogen containing amino, which is part of the amino acid, is removed and changed into urea, which is later excreted by the kidneys as urine.

3. Storage of Iron

Millions of red blood cells breakdown every day in the body. The iron which they release from their hemoglobin is stored in the liver.

4. De-toxication

Most poisonous compounds, which are absorbed in the intestine, are made harmless within the blood, that passes through the liver on its way to the general circulation.

5. Heat Production

All these chemical changes going on in the liver, release heat as a by-product. The heat is then distributed around the body by blood circulation, which helps to keep up the blood temperature.

GLOSSARY

ABSORBED

Means to take in.

ACIDS

A sharp tasting chemical, often in liquid form.

ADOPTION

To adjust to new circumstance, or surroundings.

ALKALINE

Any base like, that is soluble in water, and gives off ions in solution.

ALIMENTARY CANAL

The tube, running from the mouth to the anus. Digestion and absorption take place here.

ANAL SPHINCTER

The release of feces from the rectum, is controlled voluntarily by the anal sphincter.

AMINO ACIDS

Any of the nitrogenous organic acids, that form proteins necessary for all life.

AMYLASE

A starch digesting enzyme.

APPENDIX

A small, saw like appendage of the lower right portion of the large intestine.

ARTERY

A blood vessel, that transports blood away from the heart.

ASCENDING COLON

Part of the large intestine, that goes up vertically on the right side of the body.

ASCENDING

Sloping upwards.

ATOM

The smallest part of an element.

BACTERIA
Micro-organisms,which have no chlorophyll, and multiply by simple
 division. They are harmful and useful. Useful for
 fermentation harmful causes diseases.

BASE
A substance, that forms salt, when it reacts with an acid.

BILE DUCT
The pipe through which, the bile pass from the liver.

BLOOD
The red fluid, circulating in the arteries and veins.

BLADDER
A bag, that can expand in size.

BLOOD VESSEL
An artery, vein or capillary.

BLOOD PLASMA
A liquid component of the blood.

BLOOD CAPILLARY
Tiny blood vessels, connecting the arteries with veins, through which the
blood passes.

BOWELS
An intestines of a human being.

BOLUS
A round mass, especially a moist mass of food in the mouth, and prepared
for swallowing.

BRAIN
The coordinating center of the nervous system, which in vertebrates consists of a highly organized mass of tissues, situated at the anterior end of
the spinal cord, and enclosed in the bony cranium.

CAECUM
A blindly ending sac, at the junction of the small, and large intestines.

CANINES
Any of the sharp pointed teeth next to the incisors.

CAPILLARY
Any of the tiny blood vessels, connecting the arteries with veins. Fluid part of circulating blood, consisting of 91% water and 9% solids.

CAPILLARIES
The smallest type of blood vessel in the circulatory system.

CARBON
A black substance, that is formed in charcoal, or soot.

CARDIAC SPHINCTER
The opening, where the esophagus meets the stomach. It relaxes and allows food to enter the stomach, and contracts to stop food from coming up out of the stomach.

CARBOHYDRATE
An organic compound composed of carbon , hydrogen, and oxygen as a sugar or starch.

CARBON DIOXIDE
A heavy, colorless, orderless gas. It passes out, if the lungs is
 in respiration.

CAVITY
Hollow place in the tooth.

CECUM
A cavity, that opens at the end.

CELLULOSE
The chief substance in the cell walls.

CELLS
The fundamental unit, of which all organisms are composed.

CELL DIVISION
The process, by which two new cells are formed, from a single parent cell.

CELL MEMBRANE
The surface layer of cells.

CEMENT
The white hardened part of the teeth.

CHEMICAL
Any substance made by a chemical process.

CHEMICAL DIGESTION
A digestion process, that takes place in the stomach, and the intestines through chemical process.

CHEWING
Biting or crushing with the teeth.

CHYMOTRYPSIN
Proteinase digestive enzyme, found on pancreatic juice, formed in the pancreas.

CIRCULATION
The movement of the blood through the arteries, and veins.

COLON
Part of the large intestine, from the caecum to the rectum, or the first part of the large intestine, where water is absorbed.

COMPOUNDS
Two or more elements, joined together to form a compound.

CONCENTRATION
An increase in density, or a proportion of molecules in a unit volume.

CONTRACTION
Becoming smaller.

CORROSIVE
To eat away.

CROWN
The exposed part of the tooth, the crest, the head of the grinding surface of the tooth.

DE-AMINATION
Removal of the nitrogenous part of an acid, prior to converting it to glycogen.

DEFECTION
To excrete waste matter from the bowls.

DEGENERATE
To lose quality.

DENTIN
The hard tissue, under the enamel of a tooth.

DELICATE
Not Strong.

DESCENDING
Directed downwards.

DIABETES
The disease, that is caused when the blood contains too much sugar, and the pancreas is not producing enough insulin.

DIAGRAM
A drawing, that explains something, as by outlining its parts.

DIGESTIBLE
That which can be digested.

DIGESTION
The process of breaking down food particles, so that they can be absorbed into he body.

DIGESTIVE ENZYMES
These are enzymes that accelerates the rate at which, insoluble compounds are broken into soluble ones. These are: amylase which act on starch, proteins which act on protein, and lipase which act on fat.

DIGESTIVE GLAND
Part of the intestine which produces digestive enzymes and in which food is digested.

DILATION
Process of widening.

DISSOLVE
To reduce into smaller particle of a substance which when mixed with water and disappears in it, is said to have dissolved.

DUCT
A tube, or a channel for passage of a liquid substance or both.

DUODENUM
The first part of the small intestine, opening from the stomach.
EGESTION
The substance, which cannot be broken down into smaller particles.
ENAMEL
The hard, white coating of teeth.
ENTEROKIN
Enzyme produced by duodenal cells, that activates trypsinogen, by cleavage of a peptide bond to produce trypsin.
ENZYME
Helps in the digestion, by breaking down food particles, that can be absorbed into the body.
EPIGLOTTIS
The thin cartilage lid, that covers the windpipe during swallowing.
EPITHELIUM
A layer of cells in an animal, lining the inside of certain organs.
ESOPHAGUS
The gullet, a tube conveying food from mouth to the stomach.
EXCRETE
To take away waste material, or waste products of chemical reactions in the cells of the body.
FECES
The undigested material, or waste products of chemical reactions in the cells of the body.
FAT
An oily or greasy material, found in animal tissue, and plant seeds.
FATTY ACIDS
Any group of organic acids in animal, or vegetable fats and oils or organic acids, containing carbon, hydrogen and oxygen only.
FERMENTATION
The breakdown of food material by yeast or bacteria, to produce energy plus carbon dioxide, and in some cases, alcohol.

FIBERS
A thread like structures, that combines with others, to form animal or vegetable tissue.

FLUID
That can flow, and change rapidly, and easily.

FOOD
Any substance, taken in by a plant or animal, to enable it to live and grow.

FOOD PIPE
The tube, through which the food passes through to the stomach.

FUNDUS
The base of an organ, or the base of the stomach.

GALL BLADDER
A sac, that contains a liquid like substance, that tastes bitter.

GASTRIC JUICE
The clear, acid digestive fluid, produced by glands in
the stomach lining.

GLANDS
Any organs, that separates certain elements from the blood, and secretes them for the body use or throw off.

GLUCOSE
A simple sugar, or kind of sugar, that is found in all living cells.

GLYCOGEN
A substance in animal tissues, that is changed into glucose as the body needs it. The liver changes glycogen to glucose, and releases it into the bloodstream, to maintain healthy blood sugar level. Glucose is the body's best sugar.

GUM
The firm flesh, surrounding the base of the teeth.

HABITAT
A place in which a plant or animal lives.

HEART
Hollow muscular organ, which by rhythmic contractions, pumps blood round the body.

HEMOGLOBIN
The red, iron-containing pigment in the red blood cells, it can combine with oxygen.

HEMOPHILIA
A disease, in which the blood fails to clot.

HORMONE
A substance produced by endocrine glands into the circulatory system, that regulates the rate of bodily activities.

HYDROCHLORIC ACID
A strong, highly corrosive acid,which is a water solution of the gas, an d hydrogen chloride.

HYDROGEN
A gas, which burns rapidly, or a chemical secreted onto the blood stream on one part of the body, that controls the activity of other parts.

HYDROGEN CHLORIDE
A digestive juice, that controls the level of acid inside the stomach.

HYDROLYSIS
The addition of the hydrogen and hydroxyl ions of water to molecules.

ILEUM
The major part of the small intestine.

IMPULSES
Is a wave of motion.

INCISORS
Any of the front cutting teeth, between the canines.

INORGANIC
These are the substances, that do not have to come from a living organism. e.g. iron, salt, carbon dioxide, oxygen, etc.

INGESTION
To take food into the body.

INTESTINE
The lower part of the alimentary, extending from the stomach to the anus, and consisting of a long winding upper part (small intestine), and a shorter, thicker lower part (large intestine) bowels.

INTERNAL STRUCTURE
The inside arrangement.

INTERVALS
Once in a while, or a time in between.

JAW
Two bony parts, that hold the teeth and frame of the mouth, or two movable parts, that grasp or crush something. It bears teeth.

JEJUNUM
Part of the small intestine between duodenum and ileum in mammals. It has large villi, and is the main absorptive region.

JUICE
A liquid in or from animal tissue.

KIDNEY
A pair of organs, that separates waste products from the blood, and excretes them as urine.

LACTEAL
The tube in the center of a villus, into which passes the products of fat digestion in the intestine.

LACTIC ACID
A substance, that is produced in the breakdown of glucose during respiration.

LACTOSE
A sugar present in milk.

LARGE INTESTINE
The intestine of vertebrates, which include the caecum, colon and rectum.

LARYNX
The structure at the upper end of the trachea, containing the vocal. cords. The upper part of windpipe, which communicates with the pharynx.

LARYNGOTRACHEAL
Larynx and trachea. Chamber into which lungs open in amphibian.

LIPASE
An enzyme, which breaks down fat into fatty acids.

LIVER
The largest organ in the vertebrae animals.

LONGITUDINALLY
A diagram, placed lengthwise. Or along the length.

LUBRICATE.
To make it slippery.

LYMPH
A whitish fluid, that is derived from the blood plasma, and retuned to the circulation, via the lymphatic system.

LYMPH NODES
Where the white blood cells are produced .

LYMPHATIC
A vessel, which returns lymph from tissues, to the circulatory system.

LYMPHATIC SYSTEM
Network of fine capillaries, extending throughout the body
in the vertebrates, connected at points to the blood circulatory system.

LYMPHOCYTE
A white blood cell, which produces antibodies.

MALTOSE
Sugar produce, when starch is broken down by enzyme action.

MEMBRANE
A thin, soft layer of animal or plant tissue, that covers
or lines an organ, part, etc.

METABOLISM
Integrated network of biochemical reactions in living organisms.

MOLARS
The teeth used for grinding.

MOLECULE

Smallest particle of a substance.

MUSCLE
Contractile animal tissue involved in movement of the organism, and which also forms part of many internal organs.

MUCOUS
Secretion containing mucus.

MUSCULARISEXTERNA
Layer of the gut wall between the sub-mucosa and serosa, consisting of a sheet of longitudinal, and sheet of circular muscles.

MUCUS
A sticky fluid, produced by animals and humans to lubricate and protect delicate surfaces, mainly stomach wall. It dilutes Hydrochloric acid inside the stomach.

NASAL
To nose.

NECK
Point of connection.

NERVES
Bundle of fibers, which carries impulses away from the cell body of neuron.

NEURON
Nerve cell, basic unit of the nervous system, specialized for the conveyance and transmission of electrical impulses.

NEUTRALIZED
That has no effect, or function.

NITROGEN
Naturally occurring.

NITROGENOUS
Containing nitrogen.

ORGAN
A group of tissues, working together to do a particular job.

ORGANIC

This commonly refers to a substance, produced by a living organism. Organic chemicals are things, like carbohydrates, protein, and fat. They're often insoluble in water.

OSMOREGULATION
The control of the quantity of water, entering and leaving the cells of an organism.

OXIDIZED
To unite with oxygen, as in burning.

PANCREAS
The gland beneath the stomach, which secrets digestive enzymes into the smallest intestine. It also produces the hormone, insulin.

PANCREATIC JUICE
A secretion containing digestive enzymes.

PANCREATIC DUCT
Carries enzymatic fluid from the pancreas.

PAROTID GLAND
Pair of salivary glands, opening into the mouth in some mammals.

PARTICLES
Tiny bits, either suspending in liquid, or in air.

PEPSIN
A stomach enzyme, aiding in the digestion of proteins.

PEPTIDES
A chemical consisting of a chin of amino acids and resulting form the partial digestion of a protein.

PEPTIDE
A chemical, consisting of a chain of amino acids, and resulting from the partial digestion of a protein.

PEPTONES
Polypeptide product of hydrolysis of proteins by enzymes such as pepsin.

PERISTALSIS
The contractions and dilation of the alimentary canal, moving the food ball downward to the stomach.

PERISTALTIC WAVE
The to and fro movement of the food along the esophagus. The alternative contraction and relaxation of the muscles, closing behind the swallowed food, and opens in front of it, combined to move liquid and solid food along the digestive tract.

PHARYNX
The area at the back of the mouth cavity, leading into the nasal cavity, gullet, and windpipe. The throat in humans, and other vertebrates.

PHAGOCYTE
A white blood cell, which can ingest foreign particles inside the body.

PIGMENT
Any coloring matter in the tissue of plants, and animals.

PLASMA
Accounts 55% of the volume of normal blood.

PLASMOLYSIS
Partial collapse of a cell, as a result of withdrawal of water by osmosis.

POISON
Substance, which in small quantity, can cause illness or death.

POISONOUS
A substance, that can injure, or kill as poison.

POLYPEPTIDE
A chain of amino acids, linked together by peptide bond.

PREMOLAR
The first molars in the teeth.

PRODUCTION
Act of producing something, or make something out.

PROJECTION
A pointed out finger-like object.

PROTEIN
This is a substance, that occurs in living matter and is essential for diet. Chemicals with large molecules, containing carbon, hydrogen, oxygen, and nitrogen.

PROTEINASE
An enzyme, which breaks down proteins.

PTYALIN
Is an enzyme in saliva, that breaks down the starch into simpler form, known as simple sugars.

PULP
Soft, sensitive tissue in the centers of a tooth.

PULP CAVITY
Is a hollow space, inside the tooth.

PYLORUS
The opening from the stomach into the duodenum.

PYLORIC SPHINCTER
The connection between the stomach, and the small intestine.

RENIN
It turns milk into curd, and they are further act on and dismantled by pepsin.

RESIDUE
That which is left, after part is taken away or the remainder.

ROOT
The embedded part of a tooth, or a hair.

SALIVA
The watery fluid, discharged by glands in the mouth, and it aids in digestion.

SALIVARY
Glands opening into or near the mouth which secrete saliva.

SALIVARY GLANDS
Glands that produce saliva in the mouth.

SECRITIN
A hormone produced by the lining of the duodenum when the acid contents of the stomach reach it. It stimulates the pancreas to produce enzymes.

SECRETION
Material or fluid which is produced and released from a cell or gland.
SEROSA
Fine membrane over external organs.
SMALL INTESTINE
The narrow section of the intestine, extending from the stomach to the large intestine. It is divided into 3 parts: duodenum, jejunum, and ileum.
SOLUBLE
That which can be dissolved in water.
STARCH
The large pouch of the intestine, between the esophagus and intestines in vertebrates.
SPHINCTER
A circle of muscle in a tube or duct. Up and down of the stomach. It regulates the movement of food through the digestive tube. To prevent backflow of partially digested food.
SUBLINGUAL
Beneath the tongue.
SUB MAXILLARY
Beneath the lower jaw.
SUB-MUCOSA
Layer of gut wall, between the mucosa and external muscular.
SWALLOWING
The act of passing from the mouth into stomach.
TEMPERATURE
The degree of hotness or coldness of anything.
TISSUE
A number of cells, that look the same in the structure and do the same function.
TONGUE
The movable, muscular structure in the mouth, used in eating,

tasting, and in humans, the tongue.

TRANSVERSE
Lying across of a cross section.

TRYPSINOGEN
Digestive enzyme, found in the pancreatic juice of mammals.

UREA
A nitrogen that has chemical, and is formed in the liver from
excess amino acids.

URETER
The tube along which, the urine passes from the kidney to the bladder.

URINE
A mixture of water, salt, and urea, removed from
the blood by the kidneys.

VACUOLE
1. A fluid cavity, in the center of a plant cell.
2. Droplets of fluid, in the cytoplasm of animal cells.
3. Large cavity in a cell.

VEINS
Blood vessels, that carry blood to the heart.

VESTIGIAL
A degenerate part, more fully developed in an earlier stage.

VILLI
Finger-like projections, from the internal surface of the small intestine.
Finger-like objects.

VILLUS
One of thousands of Finger-like protrusions, from the internal surface of
the small intestine.

TEST

OBJECTIVE TEST ONE

1.The process of taking food into the mouth is called :

A. sleeping
B. eating
C. sitting
D. Biology
E. chewing

2.The food is swallowed down to the:

A. head
B. legs
C. stomach
D. hands
E. Earth

3.Food is first broken into before it will be absorbed by the body:

A. particles
B. blocks
C. Blues
D. clors
E. stars

4.When the food is dissolved, it is absorbed into:

A. head
B. stomach
C. liver
D. blood stream
E. kidneys

5.The blood takes the broken down food to all parts of the body except:

A. muscles
B. brain
C. heart
D. kidneys

E. Gunjur

6.The tube that runs from the mouth to the anus is called:

A. Highway
B. Africa
C. Alimentary Canal
D. Kidney
E. Ear

7.Along the Alimentary canal, food is broken down and absorbed by:

A. body
B. brain
C. heart
D. Kidneys
E. Gunjur

8.A layer of cells in the Alimentary canal is called:

A. liver
B. epithelium
C. nose
D. mouth
E. muscles

9.What is the use of mucus in the alimentary canal:

A. Lubrication
B. Cementing
C. Wasting
D. painting
E. Block

10.Alimentary canal can act as:

A. Kidney
B. Ear
C. Legs
D. Protective coating
E. Tongue

11.Mucus enzymes are transported through the tubes called:

A. Ducts
B. hands
C. hairs
D. Blues
E. boxes

12.Which one of the following glands produce mucus enzymes:

A. Pancreas and salivary
B. Hand and Ears
C. Ears and nose
D. Head and body

13.The movement of food down the food pipe is called:

A. runner
B. peristalsis
C. heating
D. rolling
E. down

14.What action pushes the food down to the stomach:

A. waves of contractions
B. Sitting
C. running
D. talking
E. sleep

Use this diagram to answer questions 15 to 18.

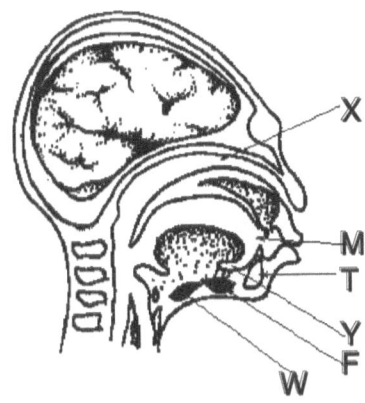

15.The X in the above diagram indicates:

A. head cavity
B. nasal cavity
C. hand cavity
D. mouth cavity

16.The arrow M in the diagram indicates:

A. mouth cavity
B. parotid gland
C. tongue
D. gland
E. sublingual gland

17.The arrow F in the diagram indicates:

A. parotid gland
B. submaxillary gland
C. nasal cavity
D. sublingual gland
E. mouth

18.The arrow Y in the diagram indicates:

A. nose
B. tongue
C. parotid gland
D. mouth cavity
E. submaxillary gland

Use this diagram to answer questions 19 to 20.

19.This diagram on the right represents:

A. Food pipe
B. A stick
C. A bar
D. muscle
E. A rocket

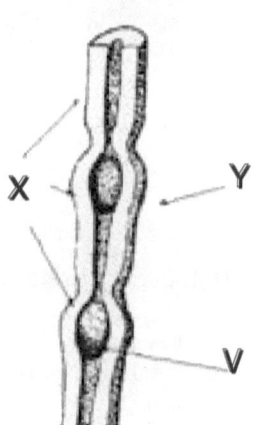

20.X and V in the diagram represent respectively:

A. waves of contraction and food box
B. waves of expansion and food ball
C. waves of contraction and food ball
D. waves of contraction and basket ball
E. waves of contraction and football

OBJECTIVE TEST TWO

1.The process of breaking down of large pieces of food into smaller particles
that can be absorbed is called:

A. digestion
B. sleeping
C. ingestion
D. ejection
E. clotting

2.This chemical helps in the breakdown of food:

A. Hydrogen

B. lead
C. enzyme
D. bond
E. glucose

3.Enzymes are:

A. chemicals
B. woods
C. stones
D. sticks
E. breads

4.Starch in food can be broken down into:

A. glucose
B. blood
C. yam
D. glycerol
E. air

5.Proteins in beans are changed into smaller soluble substance called:

A. Amino acids
B. proteins
C. corns
D. enzymes
E. tar

6.Fat molecules can be changed into:

A. blood and water
B. fatty acids and glyarol
C. glycerol and fatty acids
D. liver and salt
E. Proteins and eggs

7.Breakdown of food inside alimentary canal takes place in:

A. head
B. hands

C. different parts of the alimentary canal
D. appendix
E. liver

Use this diagram to answer questions 8 to 12.

8.The arrow G indicates:

A. Serosa
B. Mucosa
C. outside wall
D. Submucosa
E. externa

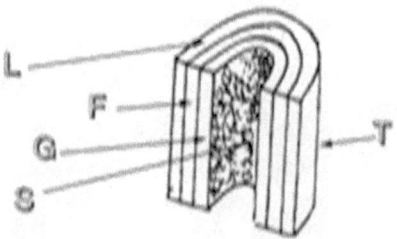

9.What does the arrow L represent?

A. Serosa
B. Mucosa
C. Outside wall
D. Food pipe
E. Food ball

10.What does the arrow T represent?

A. Outside wall
B. Muscosa
C. serosa
D. submucosa
E. waves of contraction

11.What does the arrow S represent?

A. Neck
B. Food pipe
C. Waves of contractions

D. mucosa

E. outside wall

12.What does the arrow F represent?

A. Muscularsis
B. Muscosa
C. submucosa
D. outside wall
E. food ball

13.The process of taking food into the mouth is called:

A. ingestion
B. gland
C. digestion
D. ducts
E. saliva

14. Saliva is a:

A. blood
B. salt
C. digestive juice
D. eating
E. meat

15.Saliva is excreted from the :

A. blood
B. glands
C. tail
D. head
E. ear

16.This gland is situated at each side of the lower jaw:

A. parotid gland
B. submaxillary gland
C. sublingual gland
D. enzymes

E. mouth

17.This gland is found underneath the tongue:

A. sublingualar gland
B. parotid gland
C. submaxillary gland
D. mouth gland
E. leg gland

18.This gland is the largest gland in the mouth:

A. sublingual gland
B. submaxillary gland
C. parotid gland
D. ear gland
E. liver gland

19.The food travels through this into the stomach:

A. esophogus
B. liver
C. ear
D. nose
E. lung

20.One of the following is not a gland:

A. submaxillary
B. sublingual
C. parotid
D. fundus

TEST THREE

1.When the food goes inside the stomach it does what?

A. expands
B. contracts

C. hollowed
D. small
E. medium

2.Pyloric sphincter is situated at:

A. middle of the stomach
B. end of the stomach
C. inside of the stomach
D. base of the stomach
E. top of the stomach

3.Phyloric sphincter is a:

A. bone
B. paper
C. muscle
D. car
E. meat

4.Cardiac sphincter can be found at:

A. upper end of the stomach
B. base of the stomach
C. middle of the stomach
D. outside bottom of the stomach
E. Inside end and bottom

5.What is the name of the juice that the stomach wall produces:

A. gastric
B. blood
C. liquid
D. white blood
E. pour

6.What is the name of the enzyme that the gastric juice contains:

A. pepsinogen
B. peptoid
C. pepsin

D. parotid
E. shell

7. In the stomach, the proteins are broken down into:

A. peptides and peptanes
B. blue and green
C. yellow and black
D. white blood cells
E. muscles

8.The stomach will produce an acid called:

A. Hydrochloric acid
B. Base acid
C. Protein acid
D. Blood acid
E. Peptons

9.The time that the food spends in the stomach depends on :

A. the type of food
B. fast cooked
C. raw level of the food
D. the size of the food
E. the depth of the food

10.Whay types of food spend the longest time in the stomach:

A. food that contains protein and fat
B. food that contains water
C. food that contains sugar
D. food that is small
E. big food

Use this diagram to answer questions 11 to 17.

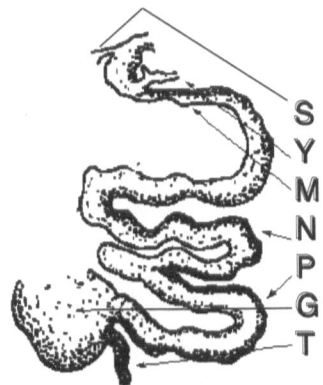

S
Y
M
N
P
G
T

11.The arrows G and T represent:

A. the small intestine
B. the large intestine
C. parts of the large intestine
D. ileum
E. caecum

12.What does the arrow S represent?

A. Bile duct
B. duodenum
C. Pancreas duct
D. Appendix
E. ileum

13.What does the arrow T represent?

A. caecum
B. appendix
C. duodenum
D. jejunum
E. pancreatic duct

14.What does the arrow Y represent?

A. bile
B. kidney
C. ileum
D. liver
E. pancreatic duct

15.What does the arrow N represent?

A. jejunum
B. ileum
C. appendix
D. bile duct
E. kidney

16.What does the arrow P represent?

A. liver
B. ileum
C. kidney
D. appendix
E. mouth

17.What does the arrow G represent?

A. Bile
B. caecum
C. ileum
D. large intestine
E. kidney

18.The pancreas is a:

A. digestive gland
B. meat
C. bar
D. stomach base
E. liver

19.Pancreas does not produce one of the following:

A. trypsin
B. amalyse
C. lipase
D. liver

20.Which one of the following act on fat:

A. lipase
B. amalyse
C. trypsin
D. water
E. ear

TEST FOUR

1.What enzyme changes starch to maltose?

A. amalyse
B. lipase
C. protein
D. trypsin
E. bloat

2.This enzyme breaks down proteins and peptides into amino acids:

A. lipase
B. trypsin
C. amalyse
D. protein
E. white cell

3.The digestive juice that comes from the pancreas is called:

A. pancreatic source
B. pancreatic juice
C. protein
D. white blood cells
E. parotid

4.Bile is produced by:

A. the liver
B. kidney
C. ear
D. hand
E. legs

5.The pancreas does not work well in:

A. acidic conditions
B. warm enviroments
C. big space
D. small space
E. yellow area

6.What does bile contain that acts of fats:

A. sugar
B. sugar
C. milk
D. bread
E. sour

Use this diagram to answer questions 7 to 10.

7.What does the arrow Y represent?

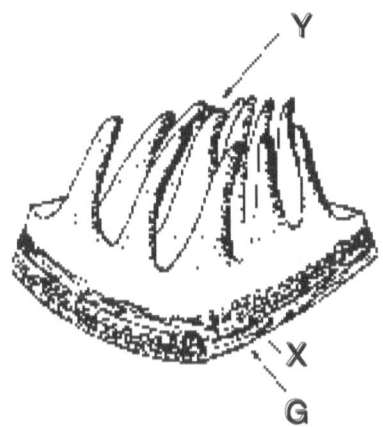

A. villus
B. villi
C. muscle
D. neck
E. fingers

8.What does the arrow X represent?

A. villus
B. circular muscle
C. longitudinal muscle
D. villi
E. neck

9.What does the arrow G represent?

A. longitudinal muscle
B. gland
C. villus
D. villi
E. circular muscle

10.What does the arrow X represent?

A. circular muscle
B. villi
C. villus
D. finger
E. mouth

11.The final products of all digestible materials are changed to:

A. soluble substances
B. hard substances
C. warm particles
D. long organs

E. beef steaks

12. All the followings are the final products of digestion except:

A. liver
B. food
C. starch
D. proteins
E. fats

Use this diagram to answer questions 13 to 19.

13. What does the arrow T represent?

A. mule
B. lacteal
C. epithelium
D. liver
E. lymphatic system

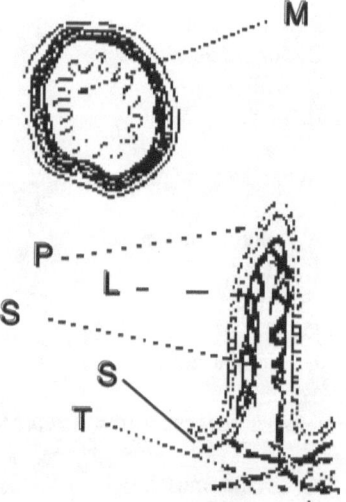

14. What does the arrow P represent?

A. lacteal
B. blood
C. epithelium
D. mouth
E. hands

15. What does the arrow L represent?

A. liver
B. blood
C. blood vessel
D. blood capillary network
E. lacteal

16. What does the arrow S represent?

A. blood vessel to villus
B. lymphatic system
C. toe

D. lacteal

E. blood tubes

17.What does the arrow M represent?

A. epithelium

B. lacteal

C. lymphatic system

D. blood vessel to villus

E. villus

18.What do the arrows M, P and T represent?

A. epithelium, lacteal, lymphatic system

B. lacteal and epithelium

C. blood vessels to villus

D. lymphatic system, liver, and kidney

E. villus, epithelium, and lymphatic system

19.What does the arrow L and Q represent?

A. lacteal veins

B. blood and lacteal

C. epithelium and lacteal

D. blood capillary network and lacteal

20.When veins come together and form one large vein is called:

A. the hapatic portal vein

B. blood vein

C. villus

D. villi

E. hepatic

TEST FIVE

1.When the digested food is replaced from the liver, where does it
enter?

A. the general blood circulation
B. abdomen
C. hands
D. proteins
E. parotid

2.Where does the fluid in the lacteal flow?

A. lymphatic system
B. proteins system
C. gland system
D. ear system
E. nose system

3.Some of the minerals that are not digested in the small intestine, go to:

A. liver
B. mouth
C. large intestine
D. sides
E. chest

Use this diagram to answer questions 4 to 9.

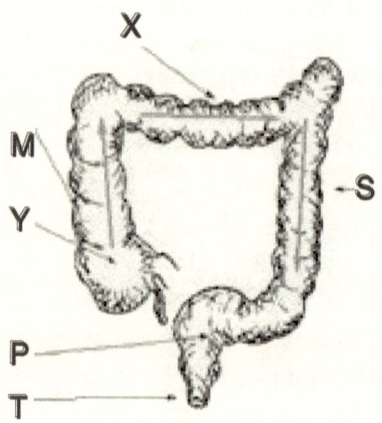

4.What does the arrow S represent?

A. Ascending colon
B. descending colon
C. anus
D. toe
E. caecum

5.What does the arrow X represent?

A. transverse colon
B. caecum
C. rectum
D. anus
E. hot

6.What does the arrow T represent?

A. rectum
B. anus
C. ascending colon
D. descending colon
E. caecum

7.What does the arrow Y represent?

A. caecum
B. kidney
C. ascending colon
D. transverse colon
E. anus

8.What does the arrow P represent?

A. rectum
B. caecum
C. descending colon
D. liver
E. kidney

9.What does the arrow M represent?

A. ascending colon
B. caecum
C. anus
D. rectum
E. descending colon

10.How many enzymes does the large intestine produce?

A. no enzymes
B. one
C. two
D. three
E. four

11.Bacteria can be found in one of the following:

A. colon
B. small intestine
C. blood
D. light
E. fire

12.What is the name of the body that had no use in humans:

A. appendix
B. ear
C. nose
D. head
E. air

Use this diagram to answer questions 13 to 20.

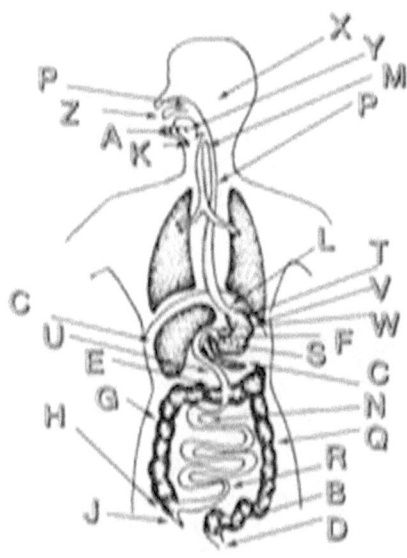

13. What does the arrow J represent?

   A. caecum
   B. appendix
   C. liver
   D. anus
   E. pylorus

14. What does the arrow L represent?

   A. liver
   B. kidney
   C. cardiac sphincter
   D. caecum
   E. pancreas

15.What does the arrow C represent?

A. liver
B. ascending colon
C. fundus
D. stomach
E. ileum

16.What does the arrow E represent?

A. mouth
B. head cavity
C. nasal cavity
D. pharynx
E. transverse colon

17.What does the arrow B represent?

A. parotid gland
B. mouth cavity
C. liver
D. nasal cavity
E. rectum

18.What does the arrow V represent?

A. liver
B. head
C. stomach
D. esophogus
E. ileum

19.What does the arrow A represent?

A. parotid gland
B. submaxillary gland
C. kidney
D. liver
E. fundus

20.What does the arrow M represent?

A. mouth
B. rope
C. esophogus
D. caecum
E. appendix

TEST SIX

1. During respiration in the cells, glucose is oxidized to form what?

A. blood and sugar
B. carbon dioxide and water
C. C2O and H4O
D. proteins
E. fats

2.One of the following is a product of digestion:

A. fats
B. woods
C. liver
D. blood
E. starch

3.Fats can be found in one of the following:

A. computers
B. cell memebranes
C. belows
D. sticks
E. cements

4.Amino acids can be built into what?

A. proteins
B. blood
C. parotid

D. meat
E. steel

Use this diagram to answer questions 5 to 9.

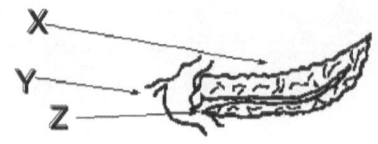

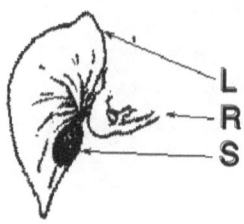

5.What does the arrow Z represent?

A. liver
B. pancreatic duct
C. pancreas
D. bile
E. neck

6.What does the arrow X represents?

A. pancreas
B. duct
C. neck

D. hand
E. liver

7.What does the arrow S represent?

A. pancreas
B. meat
C. gall bladder
D. teeth
E. hands

8.What does the arrow L represent?

A. liver
B. gall bladder
C. kidney
D. mouth
E. hands

9.What does the arrow Y represent?

A. to the liver and gall bladder
B. pancreatic duct
C. liver
D. neck
E. blood vessels

10.Which of the following can represent the natural color of the liver?

A. blue
B. green
C. red
D. yellow
E. gray

11.All the blood from the blood vessels of the alimentary canal can pass through what organ?

A. kidney
B. liver

C. stomach
D. neck
E. hair

12.After eating, what excess does the liver removes from the blood:

A. starch
B. glucose
C. parotid
D. pancreas
E. liver

13.What organ in the body regulates the blood sugar?

A. liver
B. kidney
C. liver
D. tongue
E. lips

14.The iron is stored in one of the following:

A. liver
B. kidney
C. metals
D. steel
E. stomach

15.The level of blood temperature can be controlled by one of the following:

A. heat
B. air
C. current
D. fire
E. motor

16.Heat is released in the body as a result of one of the following:

A. chemical changes
B. chemical weight

C. sitting
D. laughing
E. looking

17.Nitrogen containing amino acid can be changed into one of the following:

A. water and milk
B. urea
C. blood
D. powder
E. rice

18.Most poisonous compounds are absorbed into one of the following:

A. liver
B. kidney
C. intestine
D. stomach
E. nose

19.Liver has an ability to change glycogen into one of the following:

A. glucose
B. blood
C. fluid
D. H2O
E. parotid

20.Where is the liver situated at in the alimentary canal?

A. middle
B. end
C. below
D. top
E. above the stomach